contents

meet the quantum

In its simplest form, quantum physics is the study of matter and radiation at an atomic level, where things work very differently from "our" world. A good example of this difference can be seen in the two broad types of phenomena that we are used to: particles and waves. Particles exist in one place at a time. Waves, such as sound waves, are spread out in space. At the atomic level, this distinction does not apply. Electrons, which are thought of as particles, can behave like waves. Similarly, in the case of light, which used to be thought of as occurring in waves, some behaviour can only be explained if light exists in the form of particles, known as photons. This wave-particle double-nature or "duality" can only be accounted for by quantum physics. Other new quantum phenomena include the discreteness of energy, quantum tunnelling, the uncertainty principle, and the "spin" of a sub-atomic particle, all of which we are about to explore.

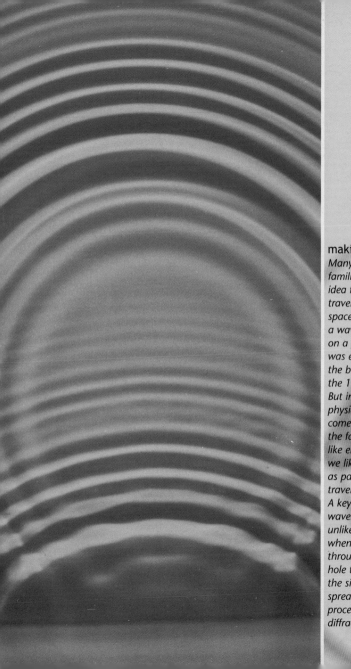

making waves

Many people are familiar with the idea that light travels through space in the form of a wave, like ripples on a pond. This was established at the beginning of the 19th century. But in quantum physics we have to come to terms with the fact that things like electrons, which we like to think of as particles, also travel as waves. A key property of waves is that, unlike particles, when they pass through a small hole they bend to the sides, and spread out, in a process called diffraction.

classical physics

Towards the end of the 19th century, physicists thought that their understanding of the way that the physical universe worked was nearly complete. Three hundred years earlier, Isaac Newton had described the laws on

Isaac Newton (1642–1727) laid the foundations of physics with his three Laws of Motion and his Law of Gravity. He also designed an improved astronomical telescope, showed that white light is made up of all the colours of the rainbow, and developed the idea that light is made up of a stream of particles, like tiny cannonballs.

which the material world operates. In 1864 James Clerk Maxwell revealed the equivalent laws controlling the behaviour of light and other electromagnetic phenomena, which apparently completed the story of how the universe of matter and light worked. Yet within a generation the world of physics was turned upside down by new discoveries of phenomena operating outside the laws of Newton and Maxwell. The "quantum revolution" that these discoveries generated is the one true revolutionary event in the whole history of science. However, to see why it was so revolutionary we need to understand precisely what was overturned by the revolution – the "classical" account of physics according to Newton and Maxwell.

Newton's laws

Among Isaac Newton's discoveries are the three Laws of Motion that describe the way all kinds of objects interact when they collide with one another in the everyday world. These three laws explain everything from the way atoms and molecules bounce off one another, how the moving parts of a car engine interact, and what's required to send a

spacecraft into orbit. Newton's First Law states that any object stays in the same place, or moves in a straight line at a constant speed, unless it is acted on by a force. The truth of this is not obvious in the everyday world because there is always a force – friction – that tends to slow down moving objects. However, objects that are moving in space in free fall, such as planets in orbit around the Sun, flawlessly obey this law because they are moving in a virtual vacuum and there is no friction.

❝I do not know what I may appear to the world; but to myself I seem to have been only like a boy playing on the seashore, and diverting myself in now and then finding a smoother pebble or a prettier shell than ordinary, whilst the great ocean of truth lay all undiscovered before me.❞

Isaac Newton

Newton's Second Law says that when a force acts on an object the acceleration produced is equal to the force applied divided by the mass of the object. This is familiar in everyday life, for example, on a pool table where the harder you hit the white ball, the faster it will move. However, acceleration is not defined solely by an increase in velocity – acceleration can also be defined as a change in direction, or by a combination of changes in both velocity and in direction. So although the Earth is moving at approximately the same speed, it is also accelerating because the Sun's gravity is constantly causing it to change

Newton rules
Newton's Laws of Motion are obeyed almost exactly on a nearly frictionless surface. The closest we can come to this on Earth is the perfectly smooth surface of an ice rink. For this reason, the sport of curling is very "Newtonian".

bouncing balls
Collisions between pool balls also obey Newton's laws. In a simple collision between two balls, if one ball is deflected to the right, the other must be deflected to the left.

direction from following a straight line (the "natural" path of any moving object) to following a curve, otherwise known as its orbit around the Sun.

The third Law of Motion is the one that causes people most trouble. Newton used the word "action" to mean "force" when he said that for every action there is an equal and opposite reaction. For example, when I fire a rifle, the action pushes the bullet out of the barrel and the reaction makes the rifle butt kick against my shoulder. If I stood on an ice rink and threw a heavy medicine ball away from me, the reaction would make me slide backwards on the ice. When a spacecraft fires its rear engine in space, exhaust gases are pushed out of the vents, producing a reaction that makes the rocket accelerate forwards. It is this law that explains how atoms bounce off one another, what happens when billiard balls collide, and why head-on car collisions can be even worse than running into a brick wall, because the combined speed of the two cars has to be taken into account.

a momentous collision
If a moving car hits a brick wall, the scattering of the bricks carries some of the momentum away. But in a head-on collision, their combined momentum is wholly absorbed by the cars, causing far more damage.

Maxwell's equations

Until the end of the 19th century, Newton's three laws seemed to define the material world completely. Even the behaviour of atoms and the newly discovered electron could be explained by a combination of Newton's laws and the application of the electric and magnetic forces.

Maxwell had provided an explanation of these two forces, the other half of the physical world, which seemed to complete the picture. Maxwell built on the research by Michael Faraday into electricity and magnetism. During the first half of the 19th century,

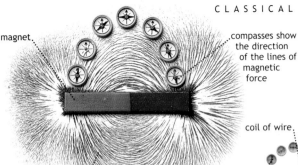

magnet

compasses show the direction of the lines of magnetic force

iron filings show the shape of the lines of magnetic force

compasses show the direction of the lines of magnetic force

coil of wire

attached leads supply an electric current

fields of force
The lines of magnetic force making up the field around a bar magnet can be traced using iron filings, or by placing magnetic compasses near the magnet. The compass needles line up along the field lines. When an electric current flows through a coil of wire, it causes a similar field of magnetic force to form, but only while the electric current is flowing.

Faraday developed the key concepts of lines of force, and magnetic and electric "fields". He invented the electric motor and the dynamo, and discovered that a changing magnetic field always produces an electric field, and a changing electric field always produces a magnetic field. A field can be thought of as the region over which the force involved has influence. It can be visualized by what happens when a bar magnet is placed under a sheet of paper covered with iron filings and the paper is gently tapped. The filings align themselves in curving patterns around the magnet, marking out the shape of its field. Each of the curved lines linking the south pole of the magnet to its north pole is a line of force.

dynamic physics
A dynamo, or electric generator, works because a magnet moving past a wire makes an electric current flow in the wire. Michael Faraday discovered that pushing a magnet into a coil of wire connected to a meter to measure electric current caused the needle of the meter to flicker.

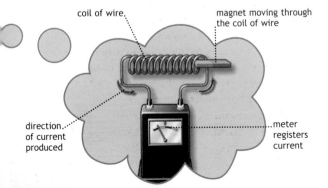

coil of wire

magnet moving through the coil of wire

direction of current produced

meter registers current

The Scottish physicist **James Clerk Maxwell** (1831–79) discovered the set of four equations that describe the behaviour of electromagnetic radiation in classical terms. This is the greatest achievement of classical physics apart from the work of Isaac Newton. As well as developing the mathematical description of light, Maxwell came up with the "three-colour" process of making coloured images, which is used in domestic TV receivers today.

Maxwell had been able to condense everything that could be said about electric and magnetic fields in a set of just four equations, which became known as Maxwell's Equations. They perform the same role in field theory (the theory of electromagnetic forces) as Newton's Laws do in mechanics (the theory of particle behaviour). Among other things, the equations describe what happens when a moving electric wave, which might be caused by an electron jiggling up and down in an atom, travels through space. As the wave ripples along, the electric field at each point it passes changes. But a changing electric field creates a magnetic field. So the electric wave creates a magnetic wave that moves in step with it. But the magnetic wave creates an electric wave. What really happens is that a jiggling electron produces not a simple electric wave, but a paired electric and magnetic wave – an electromagnetic wave. There is no such thing as a pure electric or a pure magnetic wave.

making waves
Electromagnetic waves are made when charged particles, such as electrons, jiggle about. This is analogous to the way in which a wave can be made to run along a rope by shaking the end up and down.

wavelength of the wave

wave moves horizontally along the rope.

up-and-down motion of the hand

amplitude of the wave

light as a wave

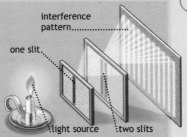

interference pattern.........

one slit...

light source   two slits

Two hundred years ago, Thomas Young demonstrated that light is a wave using an experiment with two holes. First, light with a pure colour shines through a single narrow slit in a screen. The light from this hole passes through two parallel slits in a second screen, and falls on a third screen. There, it makes a pattern of light and dark stripes. The

Thomas Young (1773–1829)

explanation is that light travels in a wave, like ripples in a pond. The ripples from each of the two slits go up and down, and where they meet they interfere. Where both ripples are in step, they make a brighter light; if they are exactly out of step they cancel out and leave a dark stripe. It was an astonishing revelation to 19th-century scientists that adding two lots of light together could make darkness.

making waves
Despite the simple structure of the equipment used by Young, the observed results were a revelation.

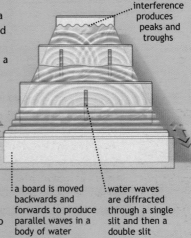

interference produces peaks and troughs

a board is moved backwards and forwards to produce parallel waves in a body of water

water waves are diffracted through a single slit and then a double slit

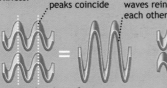

peaks coincide   waves reinforce each other

constructive interference
Peaks coincide to produce a wave of twice the amplitude.

peaks coincide with troughs

waves cancel each other out

destructive interference
Peaks coincide with troughs to produce no wave.

light enters prism

refracted light
A simple triangular glass prism will split white light (or sunlight) into the colours of the rainbow. We now know that each colour corresponds to a different band of wavelengths, and the shorter the wavelength, the greater the refraction.

refraction of white light splits it into its constituent wavelengths...................

electromagnetic spectrum
Visible light makes up just a small portion of the electromagnetic spectrum, which ranges from very short wavelength gamma radiation through to radio waves. The extent to which an electromagnetic wave is refracted depends upon its wavelength.

The icing on the cake came when Maxwell used his equations to calculate the speed at which an electromagnetic wave must travel. The equations give a unique speed for the wave, and for all electromagnetic waves. This speed turned out to be exactly the speed of light. All electromagnetic waves travel at the speed of light, and the inescapable conclusion is that light consists of an electromagnetic wave travelling through space.

light wave problems

It was already known that light travels through space in the form of a wave from the work of Thomas Young in England and Augustin Fresnel in France early in the 19th century. Faraday had suggested that light is "a high species of vibration in the lines of force", but he lacked the mathematical knowledge to describe how this could work.

Experiments like the ones carried out by Young show that visible light has a wavelength (the distance between two successive peaks of the waves) in the range from about 380 nanometres (violet) to 750 nanometres (red). The same equations were soon being used to describe longer wavelength radiation (such as radio waves) and shorter wavelength radiation (ultraviolet light and X-rays). However, there was one problem with the wave theory of light. It could not explain a phenomenon known as black body radiation. At first, this seemed to be a minor irritation that could soon be resolved, but the more physicists looked at the problem, the bigger it grew.

gamma   x-ray   ultraviolet   visible light   infrared   microwave   radio wave

the quantum revolution

Black body radiation gets its name in a curiously backwards way. In classical physics, a black body is an object that absorbs all the electromagnetic radiation that falls on it. If such an object were to become hot, it would then radiate energy – black body radiation – but it would no longer be black. Its old name – cavity radiation – gives a much more telling insight into the nature of this kind of radiation.

❝ The only laws of matter are those which our minds must fabricate, and the only laws of mind are fabricated for it by matter. ❞

James Clerk Maxwell

what is black body radiation?

Imagine a large, hollow sphere in which there is one tiny hole. Any radiation that arrives at the hole goes through and is absorbed inside the cavity, which acts like a black body. Now imagine heating the sphere up until it glows, first red-hot, then white-hot, then blue-hot. The radiation that comes out of the hole is pure black body (or cavity) radiation. The name "black body" is applied, somewhat confusingly, even to this coloured radiation. This example highlights one of the most important features of black body radiation. Its colour depends on its temperature. Since the colour of light is related to its wavelength, this means that the intensity of the radiation

radiation enters cavity

absorbing and radiating

A cold sphere with a hole in it absorbs electromagnetic radiation and acts like a black body. When it gets hot, it emits electro-magnetic radiation. This is black body radiation.

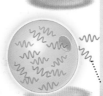

radiation escapes from cavity

emitted at each wavelength depends on the temperature of the object. This is true in everyday life, and we all know that an object like a central heating radiator may radiate infrared heat without glowing and that a red-hot lump of iron is cooler than a white-hot lump of iron. Many objects radiate very nearly like a black body, including the Sun. It is possible to measure the surface temperature of the Sun (about 6,000°K) simply from its colour (yellow-orange) and by assuming that it is a black body radiator.

By the 1890s, experiments had shown exactly how the radiation of a black body is related to its temperature. When the electromagnetic spectrum of this radiation is plotted as a graph (the black body curve), it shows a smooth peak, like a hill. At a particular temperature, the peak is always in the same part of the spectrum, that is, at the same wavelength. But, as the black body gets hotter, the peak moves to shorter wavelengths (from infrared, to red, to orange, to blue, and so on).

red-hot
A glass blower can tell the temperature of the molten glass, and therefore its physical properties, from its colour.

black body curves
The wavelength at which the intensity of the radiation from a hot object peaks is related to the temperature of the object. These "black body curves" all peak in the infrared region of the spectrum in the form of "hills", and produce little radiation in the visible spectrum.

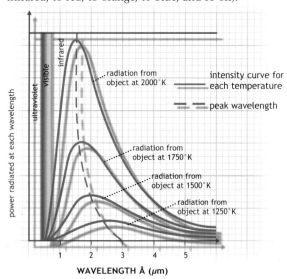

- radiation from object at 2000°K
- intensity curve for each temperature
- - - peak wavelength
- radiation from object at 1750°K
- radiation from object at 1500°K
- radiation from object at 1250°K

power radiated at each wavelength

ultraviolet | visible | infrared

WAVELENGTH Å (μm)
1   2   3   4   5

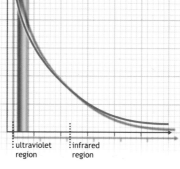

But either side of the peak there is very little radiation.

This posed a profound puzzle that could not be explained by classical physics. It goes like this. If

: ultraviolet region
: infrared region

predicted curve
This is the radiation curve that the classical physics of Newton and Maxwell predicted. Any hot object should radiate energy mainly at very short wave-lengths, which is the ultraviolet end of the spectrum. The failure of this prediction has become known as the "ultraviolet catastrophe".

electromagnetic waves are treated in the same way (mathematically speaking) as waves on the ocean or on a violin string, then if energy (heat) is put in, the intensity of the radiation produced should be proportional to the frequency of the radiation. The higher the frequency (which means the shorter the wavelength), the more radiation there should be – at any temperature. Most of the energy emitted should be in the ultraviolet region and there should be no "hill" within the graph, but there is. This "ultraviolet catastrophe" spelt the beginning of the end of classical physics as a complete explanation of the physical world.

Planck's constant

Confronted by this inexplicable phenomenon, in 1900 the German physicist Max Planck was reduced to the desperate remedy of solving the black body puzzle by assuming that light could be divided up into little chunks, pieces that he called "quanta", instead of always being a smooth, continuous wave. In order to do this, he assigned an energy, E, to each quantum that was related to its frequency, f.

Max Planck
(1858–1947) started the quantum revolution in 1900, when he explained black body radiation in terms of light quanta. Planck never really came to terms with the revolution he had started, but was open-minded enough to be one of the first people to appreciate the importance of the Special Theory of Relativity when it was published in 1905. In 1918 he received the Nobel Prize for his scientific work.

His interpretation worked provided that $E = hf$, where h was a new mathematical constant in nature, and is now known as Planck's constant in his honour.

It works like this. In any object, the energy is distributed among the atoms. A few have very little energy, a few have lots, while most have a middling amount of energy. But what we mean by "middling" changes as the temperature increases. Each atom can emit electromagnetic radiation in the form of quanta. For large values of f (high frequencies, or short wavelengths) the energy (E) needed to emit a single quantum is very large, and only a few atoms have that much energy. At low frequencies (long wavelengths) it is easy to radiate the quanta because less energy is required. However, each quantum has so little energy (because f is small) that even added together quanta do not contribute much to the spectrum. In the middle, though, there are lots of atoms producing lots of quanta, which add up to give the hill in the black body curve. Everything matched the experiments, for one particular value of h.

Planck did not think of these quanta as being like little particles, or bullets spat out by atoms. He thought there must be something about the internal workings of atoms that only allowed them to emit pulses of light, but that the light was still a wave. This is a bit like the way a cash machine works. It will only let you have money in multiples of £10, even though other amounts of money, such as £27, exist.

stepping up
Moving about in the quantum world is like moving up a step-ladder. You can only move by a whole number of steps; there is nowhere "in between" the steps on which you can rest your feet.

enter Einstein

Most of Planck's colleagues thought that his discovery, although useful, was only some kind of mathematical trick, and that when atoms were understood better it would be possible to dispense with it. At first, the only person who took and used Planck's new interpretation as it stood was Albert Einstein, then an unknown scientist just beginning his research. He

a meeting of minds
Max Planck (left) and Albert Einstein were the two originators of quantum theory, although they came from different generations.

showed in 1905 that a perplexing phenomenon at the time, known as the photoelectric effect and which also had no classical explanation, could also be accounted for and understood if light did in fact travel as a stream of minute particles. In its simplest form, the photo-electric effect is that a current can be started in a circuit by shining a light on a metal plate – electrons in the metal are agitated and create a current of electricity. The particular effects that Einstein observed were only consistent with the current being generated by the electrons being struck by particles rather than by waves. However, at first nobody else accepted the idea of these particles of light, and they were only given the name photons in 1926, after much more evidence emerged to show that Einstein was right.

> ❝ I spent ten years of my life testing that 1905 equation of Einstein's, and contrary to all my expectations, I was compelled in 1915 to assert its unambiguous verification in spite of its unreasonableness. ❞
>
> Robert Millikan (1868–1953)

light work
The photoelectric effect, explained in quantum terms by Albert Einstein in 1905, is the basis on which the solar panels on this futuristic electric vehicle convert sunlight into power.

light from the Sun generates an electric current in photoelectric cells

WORLD SOLAR CHALLENGE 16

Bohr's model of the atom

The evidence that made people take the idea of quanta seriously came from the work of the Dane, Niels Bohr. In 1911, Ernest Rutherford, from New Zealand, had discovered that the atom consists of a central, positively charged nucleus surrounded by a cloud of negatively charged electrons. It was natural to think of the electrons as somehow "in orbit" around the nucleus, rather in the same way that the planets are in orbit around the Sun. But there is a big difference. According to classical theory, an electrically charged particle orbiting in this way should continuously radiate away electromagnetic energy, and spiral in to the nucleus. What was holding atoms up?

invisible force
If electrons were literally "in orbit" around atomic nuclei, they would radiate energy and fall inwards. What holds them up?

Bohr said that electrons could only occupy certain orbits around the nucleus, each of which corresponded to a precise amount of energy which was a multiple of the basic quantum. No in-between orbits were allowed, because it was not possible to have a fraction of a quantum (just as, if we push the cash analogy to the limit, there is no physical piece of money smaller than 1p, and you can't have, say, 12.7p in your pocket). So electrons could not radiate energy continuously and spiral inwards. What they could

Danish physicist **Niels Bohr** (1885–1962) was the first to apply quantum physics to provide a description of the atom. Bohr developed his key ideas when he worked with Ernest Rutherford in Manchester between 1912 and 1916, but spent most of his career in Denmark. There, he established the Institute for Theoretical Physics (later known as the Niels Bohr Institute). He was awarded the Nobel Prize in 1922.

do was jump straight from one orbit to another, either emitting or absorbing a quantum of energy in the process.

elemental bar code
Just as the bar code on a can tells you what is inside, so the spectrum of an element, such as hydrogen, tells you which atoms are present.

The beauty of this idea was that it was already known that atoms of each element produce their own characteristic lines in the spectrum, as distinctive as a bar code. The lines correspond to a precise wavelength, which translates as a precise amount of energy. Bright lines correspond to energy radiated by an atom; dark lines to energy absorbed by an atom. The pattern of those lines made it possible to calculate the energies involved if electrons really were jumping about in this way inside atoms, and the observations matched Bohr's calculations.

HYDROGEN SPECTRUM

LITHIUM SPECTRUM

SODIUM SPECTRUM

The final ingredient in Bohr's model of the atom was his idea that an orbit only has room for a certain number of electrons. If the orbit was full, no more electrons could be added, even if there were atoms in higher orbits that would "like" to make the jump to lower energy levels. So atoms were stable, and the distinctive patterns of lines in the spectrum of light from different elements was explained. Bohr's model of the atom, presented in 1913, made people sit up and take notice of quantum physics as something more than a mathematical trick; but progress in this new branch of physics was delayed by the First World War. It was in the 1920s that ideas and conjecture came together in a fully worked-out quantum theory.

fingerprints of light
Three examples of "pure" spectra obtained from three different elements. Each kind of hot (excited) atom produces its own characteristic "bar code", so the elements producing this light can be unambiguously identified.

ff Anyone who is not shocked by quantum theory has not understood it. ff

Niels Bohr

Bohr model

Niels Bohr described the electrons associated with an atom as being in orbits, or shells, around the nucleus. These orbits are in layers, like onion rings, and the quantum rules only allow a certain number of electrons in each orbit. A full orbit is particularly stable, and atoms will combine with each other to make molecules in such a way that they share electrons to give an illusion of full orbits.

matching protons and electrons
All atoms contain an equal number of protons and electrons. Hydrogen has one proton and one electron, whereas lithium has three of each.

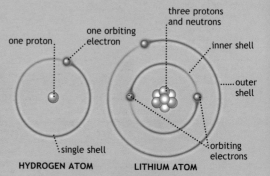

one proton
one orbiting electron
three protons and neutrons
inner shell
outer shell
single shell
orbiting electrons

HYDROGEN ATOM

LITHIUM ATOM

methane

Methane molecules are made of hydrogen and carbon. Hydrogen atoms have one electron, but would "like" to have two, to make a full shell. Atoms of carbon have a full inner shell of two electrons, and a half-full outer shell containing four electrons. So one carbon atom and four hydrogen atoms can get together to make a molecule of methane.

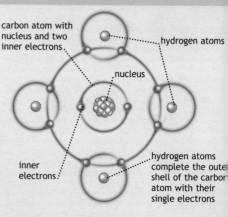

carbon atom with nucleus and two inner electrons
hydrogen atoms
nucleus
inner electrons
hydrogen atoms complete the outer shell of the carbon atom with their single electrons

METHANE MOLECULE

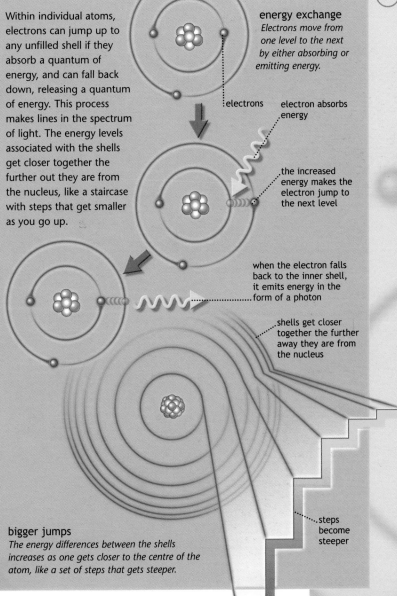

Within individual atoms, electrons can jump up to any unfilled shell if they absorb a quantum of energy, and can fall back down, releasing a quantum of energy. This process makes lines in the spectrum of light. The energy levels associated with the shells get closer together the further out they are from the nucleus, like a staircase with steps that get smaller as you go up.

energy exchange
Electrons move from one level to the next by either absorbing or emitting energy.

electrons

electron absorbs energy

the increased energy makes the electron jump to the next level

when the electron falls back to the inner shell, it emits energy in the form of a photon

shells get closer together the further away they are from the nucleus

bigger jumps
The energy differences between the shells increases as one gets closer to the centre of the atom, like a set of steps that gets steeper.

steps become steeper

out of the darkness

The double-slit experiment proved that light travels as a wave, but early 20th-century experiments showed that light is made up of particles. In the face of these irreconcilable findings, a useful resort is to develop mental models to imagine the unimaginable. One of these models revolves around waves of probability, which we look at in the opening of this section. Probability led to the idea that electrons were both particle and wave, which in turn led to the conclusion that a quantum system does not exist in a definite state of either particle or wave until it is measured, which in turn formed the basis of the notion of uncertainty. Paradoxically, these findings about the apparently impossible nature of energy and matter led to quantum electrodynamics (QED), a theory that can account for all interactions between electrons and photons, including how the Sun shines, and what holds DNA together. It has also led to the development of lasers and microchips.

diffracted waves
Monochromatic (single-wavelength) light from a laser spreads out in a diffraction pattern as it passes through an aperture, showing the wave nature of light.

the central mystery

mass and energy
The distinction between waves and particles in the everyday world is highlighted by a surfer ("particle") in a definite location riding on a spread-out wave.

One of the greatest physicists of the 20th century, Richard Feynman, described the double-slit experiment, which proved among other things that light travels as a wave, as "the central mystery" of quantum physics. There is no doubt that the experiment is correct and that it can be used to calculate the wavelength of light. Light is a wave. But in the early decades of the 20th century, repeated experiments, such as studies of the photoelectric effect, showed just as surely that light is made up of particles – photons. Many of these experiments were carried out by Robert Andrews Millikan, who opposed Einstein's interpretation of the photoelectric effect and set out to prove Einstein wrong. Ironically, his excellent work conclusively demonstrated that Einstein was correct. The evidence was far more compelling than it would have been had Millikan been trying to confirm the theory.

a probability model

background image
Because of the way waves spread out, two light waves can interfere with one another to make striped patterns.

So how can the results of the double-slit experiment be explained in particle terms? The short answer is: they can't. Nothing in our everyday experience equips us to understand what is going on. The bottom line is that the way that events occur in the quantum world is utterly different from the everyday world of our experience, and that the truth can only be fully expressed in the

Albert Einstein (1879–1955) is best known for his two theories of relativity, but his contributions to quantum theory were also profound, and he received the Nobel Prize for his work on the photoelectric effect. After graduating in 1900, he completed his Special Theory of Relativity and his work on the photoelectric effect while working as a patent officer, and he did not obtain a university post until 1909.

mathematical equations that enable physicists to predict things like the energy levels of electrons in an atom or the brightness of the stripes in a diffraction pattern.

One model that many people have found useful as an aid to understand these ideas involves waves of probability. If you threw rocks at a wall with two holes in it, the result would be two piles of rocks, one behind each hole. You would not get anything like the diffraction pattern you see when waves pass through two holes. But if the "rocks" are photons, and the holes are very small, you do get a diffraction pattern. It is as if there is a wave associated with each photon. The waves from all the photons interact and make a diffraction pattern, which can be thought of as a probability pattern in that photons will most probably end up where the waves have interfered to produce peaks in the diffraction pattern, and less likely to end up where the interference has produced a cancelling of the waves. However, there is no way to predict in advance where any individual photon will end up. This is strange enough when we think in terms of photons. But it gets even stranger when we think in terms of electrons.

strange behaviour
If two people threw a lot of rocks onto a heap, you would be astonished if they lined up in stripes like the interference patterns made by waves.

waves and particles

As the evidence mounted, light, which used to be thought of as a wave, had to be regarded as being both wave and particle. In the early 1920s, the French physicist Louis de Broglie came up with the inspired suggestion that electrons, which used to be thought of as particles, had to be regarded as being both particle and wave. In fact, he said that everything had to be regarded as being both particle and wave, but that the amount of waviness associated with everyday objects, such as people or pot plants, doesn't show up because they have so much mass.

Using Einstein's work on photons, an equation was discovered that linked the particle and wave properties of light. The equation says that the wavelength associated with a photon multiplied by its momentum (the force that keeps an object moving) is equal to Planck's constant, h. De Broglie realized that the equation is universally true, so that a particle such as an electron (or anything else) that has a certain momentum must also have a certain wavelength, which can be calculated by this equation. The wave equation describing such a quantum entity is often called the wave function. De Broglie's discovery was enthusiastically endorsed by Einstein, and experiments in which the wavelengths of electrons were measured were carried out later in the 1920s. These showed that everything in the quantum realm is both particle and wave at the same time.

subatomic sculpture
By making a ring of atoms to act as a "corral", researchers in the 1990s were able to trap electrons as waves moving across the surface of a metal like ripples on the surface of a pond.

tennis ball arrives at a definite place ········

ball spreads out into a wave form and goes through both holes at once ········

h=0.000000000000000000000000000066

m=0.0000000000000000000000000000009

However, the quantum realm is limited to the world of the very small, because Planck's constant is so small. In units where mass is measured in grams, the value of h is 6.6×10^{-27} (a decimal point followed by 26 zeroes and two sixes). Since de Broglie's equation says that the wavelength of an object is equal to this tiny number divided by its momentum, and momentum is related to mass, the wavelength will only be detectable for objects with tiny masses. Sure enough, in the same units the mass of an electron is 9×10^{-28} grams. Everything fits.

electron patterns

Although all this was clear to the experts before the end of the 1920s, the full drama of wave-particle duality and the probabilistic element that had been found in quantum physics was brought home in the 1980s by a team of Japanese scientists working at the Hitachi Research Laboratories. They carried out the double-slit experiment with electrons, which were fired, one at a time, through the equipment. The electrons were detected at a screen (like a TV screen) on the other side of the slits, with each electron making a spot of light when it arrived. The screen then "remembered" the spot, so that an image gradually built up as more and more electrons arrived.

size matters
Quantum effects are important for electrons because if you write the numbers out in the appropriate units, the mass (m) of an electron is even smaller than Planck's constant (h).

if h were larger
If Planck's constant were big enough, tennis balls would behave like electrons. After you hit them, they would dissolve into waves as they travelled from one place to another, and only condense back into particles when they hit something.

ball leaves racquet as a definite object.

the heart of the matter

Quantum entities behave as if they are both wave and particle. In an experiment in the 1980s, electrons were fired, singly, through the equivalent of Young's double-slit experiment. They were detected at a screen like a TV screen, where each one made a spot of light that the screen "remembered". Each electron was fired as a particle, and each one made one spot on the screen, confirming its arrival as a particle. But when thousands of electrons were fired through the experiment, the pattern the spots on the screen was the classic interference pattern for waves. Exactly equivalent experiments have also been done with photons (particles of light). You don't have to understand this. Nobody understands it. You just have to accept that this is the way the quantum world works.

electrons are first fired through a single slit one at a time

the electrons then interact in an unknown way with the double slits

Each of the electrons made a single spot of light, proving that they are particles. As the electrons were fired one after another, the spots of light gradually built up a clear pattern – the diffraction pattern – not the two "heaps of rocks". The electrons were not even travelling through the experiment together, but they still built up the characteristic pattern associated with waves passing through the two slits. It was as if each electron knew where all its predecessors had gone, and where all the electrons coming behind would go, as well as knowing its own place in the pattern. Nothing demonstrates the weirdness of the quantum world more clearly to non-scientists.

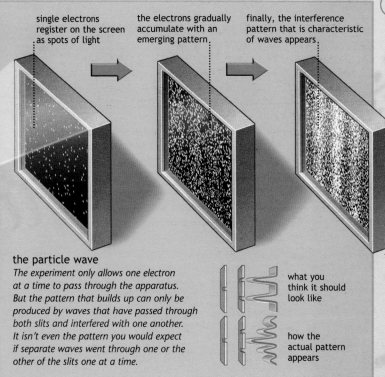

single electrons register on the screen as spots of light

the electrons gradually accumulate with an emerging pattern

finally, the interference pattern that is characteristic of waves appears

the particle wave
The experiment only allows one electron at a time to pass through the apparatus. But the pattern that builds up can only be produced by waves that have passed through both slits and interfered with one another. It isn't even the pattern you would expect if separate waves went through one or the other of the slits one at a time.

what you think it should look like

how the actual pattern appears

Of course, the electron does not "know" anything. It is simply following the blind rules of chance in the same way that a true die does not "know" what numbers have come up previously (or are coming next) in order for the probability of rolling, say, a three next time to be exactly one in six. The role of probability runs very deep in quantum physics, something Einstein hated, famously commenting "I cannot believe that God plays dice."

roll of the die
A tumbling die lands with one face uppermost in accordance with the rules of probability. It doesn't have to "know" what number came up last time. Quantum processes also obey probabilistic rules.

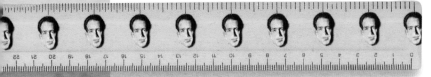

affecting reality
Niels Bohr said that what we choose to measure, and the act of measuring, affect the nature of reality. The most important feature of the Copenhagen Interpretation is that no quantum property (such as the position of an electron) is real until it is measured.

All the evidence is that Einstein was wrong and the quantum world really is ruled by probability.

The conclusions drawn from the results of experiments such as the double-slit experiments were developed at the end of the 1920s. Collectively they are known as the Copenhagen Interpretation because Niels Bohr and his colleagues in Denmark were closely involved in its development (although key ideas about probability came from Max Born in Germany).

The Interpretation says that a quantum system does not exist in a definite state until it is measured. For example, an electron travelling through the double-slit experiment does so as a spread-out wave, and does not have a precise location in space. It is only when it arrives at the detector screen that it makes a "choice" from the probabilities (like a tumbling die finally settling with three spots uppermost) which causes the wave function to "collapse", as Bohr put it, onto a single point. Using this model, quantum entities travel as waves, but arrive as particles.

Schrödinger's paradox

The idea of an electron wave function proved very useful in uniting observation and theory, and enabled the Austrian Erwin Schrödinger to develop a description of the way that electrons behave in atoms in terms of waves. This is a much more complete description than Bohr's model, and enabled researchers such as the American Linus Pauling to explain all the principles of chemistry (how atoms interact to make molecules) in terms of quantum physics. This has led physicists to quip that "chemistry is now a branch of physics".

in the balance
Pushing Bohr's ideas to their absurd limits, Erwin Schrödinger imagined a cat that would be neither alive nor dead until its state of health was "measured" in some way.

In the 1920s, Austrian physicist **Erwin Schrödinger** (1887–1961) found a wave equation that describes the behaviour of quantum entities such as electrons and other "particles". In 1933 he received the Nobel Prize for this work (jointly with Paul Dirac), but he was dismayed that the weirdness of quantum physics could not be explained in the familiar language of wave equations.

Schrödinger was particularly pleased that his wave function seemed to be bringing the common sense of familiar waves back into fundamental physics. This is why he was so horrified when he discovered that it was not possible to get rid of the probabilistic effects, and later said of his own theory, "I don't like it, and I wish I'd never had anything to do with it." He even dreamed up an imaginary "thought experiment", known today as Schrödinger's Cat Paradox (although it isn't really a paradox), to highlight what he saw as the absurdity of the Copenhagen Interpretation, which, taken literally, says that a cat can be both dead and alive at the same time.

Schrödinger failed in his attempt fully to explain electrons purely in terms of waves because it is no more true to say that an electron is a wave than to say that it is a particle. All we can say is that quantum entities such as electrons

> ❝ You believe in a God who plays dice, and I in complete law and order in a world which objectively exists, and which I, in a wildly speculative way, am trying to capture… even the great initial success of the quantum theory does not make me believe in the fundamental dice game, although I am well aware that your younger colleagues interpret this as a consequence of senility. ❞
>
> Albert Einstein, letter to Max Born (1926)

have properties (which have no counterpart in everyday life) which make them appear sometimes as particles and sometimes as waves. This is closely related to the idea of quantum uncertainty.

the uncertainty principle

The importance of uncertainty in the quantum world was first spelled out by the German, Werner Heisenberg, in 1927. He realized that, in this context, uncertainty is a precise and definite thing, implicit in the equations that describe the behaviour of things such as electrons. His analysis revealed that there are pairs of properties for

Schrödinger's cat

Erwin Schrödinger dreamed up a thought experiment to demonstrate what he saw as the absurdity of quantum physics. He asked us to imagine a cat shut up in a windowless chamber with a plentiful supply of food and other creature comforts, but also with what he called a "diabolical device" in the form of a container of poison gas linked to a sample of radioactive material. If the radioactive material decays the gas is released, and if the gas is released, the cat will die.

According to the quantum rules, it is possible to calculate a time when there is an exact 50:50 chance that the radioactive material has decayed and that the cat has survived or been killed. Quantum theory says that if we look into the box at that time, there will be a "collapse of the wave

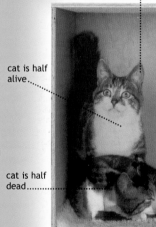

the Copenhagen Interpretation says the cat exists in a "superposition of states"

cat is half alive......

cat is half dead............

quantum entities, and it is impossible to specify precise values of both properties at the same time. The most important pair of these "conjugate variables" is position/momentum. His equations show that a particle ("quantum entity") cannot have a precisely determined position and a precisely determined momentum at the same time. And as momentum is proportional to velocity, this means that a particle does not have both position and velocity simultaneously.

In fact, the uncertainty in position multiplied by the uncertainty in momentum is always greater than Planck's constant. We do not notice Planck's constant in everyday

radioactive trigger
releases hammer

there is a
50:50 chance
that the
hammer has
fallen and
broken the
poison bottle

poison
bottle

radioactive
decay detector

radioactive
material

function" and we will see either a dead cat or a live cat. But it also says that if we don't look into the box, not just the radioactive sample but the whole experiment, including the cat, is poised in a superposition of two possible states, each placed over the other, where the cat is both dead and alive at the same time, or half dead and half alive.

Schrödinger's thought experiment highlights the fact that there is some kind of boundary between the classical world of everyday objects and the quantum world of very small objects. Nobody quite knows where that boundary is, or how quantum effects disappear when that boundary is crossed.

is Schrödinger's cat dead or alive?
This mockup of how the "cat in a box" experiment might be done shows the cat and the diabolical device poised in a superposition of states that will only be resolved if the box is opened.

Werner Heisenberg (1901–76), one of the founding fathers of quantum theory, discovered the uncertainty principle, which says that it is not not possible to measure both speed and velocity on the quantum level. He clouded his reputation by being the most able scientist to stay in Germany after the Nazis came to power and working on the German atomic bomb project during the Second World War. However, his stature had been recognized by the award of the Nobel Prize in 1932.

life because it is so small, but it is very important for an electron. The uncertainty surrounding position and momentum is not simply a result of the difficulty of measuring the position and momentum of something as small as an electron; an electron itself does not "know" precisely both where it is and where it is going.

This package of ideas (uncertainty, probability, wave-particle duality, and the collapse of the wave function), all associated with the double-slit experiment, is the essence of the nature of the quantum world. Don't worry if it makes your head hurt. Nobody understands quantum physics. What matters is that the equations associated with these ideas have many practical applications, which work whether you understand or not. We start our discussion of those applications with the jewel in the crown of quantum physics. This is a theory known as quantum electrodynamics, which is the most successful theory – in terms of its agreement with experiments carried out here on Earth – in the whole of science.

you can't know both
If Planck's constant were big enough, cars would behave like electrons. They could have a definite speed but no definite position, or a definite position but no definite speed.

Do you know how fast you were travelling?

No, but I know exactly where I am!

the jewel in the crown

The crowning achievement of quantum physics describes with great precision the way electrically charged particles interact with one another and with magnetic fields. This sounds esoteric, but it means that quantum electrodynamics, or QED, describes everything about matter that is not described by gravity. The everyday world is made up of atoms and molecules, which interact with one another through the electrons on the outer part of atoms. QED is able to account for all interactions involving electrons, and so QED is all that is needed to account for the everyday world. QED explains why the ocean is blue, how the explosions in an internal combustion engine take place, and how all chemical reactions take place. It is, in Richard Feynman's words, "the theory of light and matter" in all circumstances where gravity can be ignored. We only have to consider the other forces of nature (or "interactions", as physicists call them) when dealing with what goes on inside the nucleus of an atom – and, as we shall see, even there QED provides the archetype on which other models are based. Without the example of QED, those models would have no foundations.

at the heart of things
Quantum electrodynamics pervades every aspect of the world we see around us. It explains the blue colour of the sea, and the way chemicals react with one another.

❝All these fifty years of conscious brooding have brought me no nearer to the answer to the question, 'what are light quanta?' Nowadays every Tom, Dick, and Harry thinks he knows it, but he is mistaken.❞

Albert Einstein

The roots of QED were established in the 1930s, when two physicists, the German Hans Bethe and the Italian Enrico Fermi, suggested that interactions between charged particles could be described in terms of photons (single units of light) being exchanged between the particles. The complete version of the theory emerged in the 1940s from the work of three physicists, Sin-itiro Tomonaga in Japan, and Julian Schwinger and Richard Feynman in America. Each of the three versions that they developed is mathematically equivalent to each of the others, but Feynman's approach is easier to understand in physical terms, and has become the standard version of QED.

Feynman diagrams

The physics of QED can be understood in terms of pictures called Feynman diagrams. These represent the paths of electrons using lines, known as "world lines" that are placed within two coordinates of time and space. A vertical line indicates a stationary electron because its position in space is not changing. The more angled the world line, the greater its rate of change of position in space and therefore the faster it is moving. A photon associated with the field of a magnet can be introduced into a Feynman diagram. The photon can be shown interacting with an electron travelling in a straight line, and deflecting it onto a new path. This simple version of QED predicts that a property known as the magnetic moment of the electron has a value of exactly 1. It doesn't matter just

Richard Feynman (1918–88), the greatest physicist of his generation, was one of the founders of quantum electrodynamics. He invented the diagrams that describe the quantum world, and for 40 years he inspired young physicists with his teaching. He received the Nobel Prize in 1965 and was a key member of the panel that investigated the 1986 *Challenger* disaster.

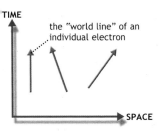

TIME

the "world line" of an individual electron

SPACE

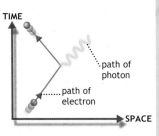

TIME

path of photon

path of electron

SPACE

getting a kick
"World lines" on a Feynman diagram represent the positions of particles in space over time. A moving electron can emit or absorb a photon, causing it to change direction.

what the magnetic moment is (you can think of it as a measure of how easily an electron can be twisted by a magnetic field), but what does matter is that it is possible to measure this property very accurately in experiments. The experiments show that the magnetic moment is actually a little bit more than one, so this simple version of QED cannot be the last word on the subject.

Feynman showed how the simple model could be improved, and drew a picture to highlight what was going on. In the next step of the calculation, while it is interacting with the photon the electron also emits and then reabsorbs another photon, interacting with itself. It is easy to draw the picture, harder to carry out the mathematical calculations. When the calculations for this interaction were carried out they

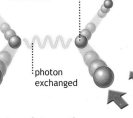

electrons deflect each other

photon exchanged

electron interacts with itself

electron interactions
Two electrons can exchange a photon, so that they repel one another. An electron can also interact with itself.

theoreticians
Seen here with Richard Feynman (right), Paul Dirac (1902–84) developed an early version of QED, and came up with the idea of antiparticles. Feynman conceived a complete theory of QED, finishing the job begun by Dirac.

predicted that the magnetic moment should be a little bit bigger than 1, though still not quite as big as the measured value, but there is no need to stop there. There is no reason why the electron cannot emit two (or more) photons, one after the other, and reabsorb them. Each extra photon allowed for in the calculations brings the calculated value of the magnetic moment ever closer to the measured value.

"There is one simplification at least. Electrons behave in this respect in exactly the same way as photons; they are both screwy, but in exactly the same way. How they behave, therefore, takes a great deal of imagination to appreciate, because we are going to describe something which is different from anything you know about."

Richard Feynman (1967)

Crucially, Feynman was able to *prove* that each time an extra photon is allowed for in the calculation, the "correction" is a little bit smaller, but always in the right direction. By the time the effect of four photons being emitted and reabsorbed was allowed for, the predicted value of the magnetic moment was 1.00115965246, while the best experimental measurement was 1.00115965221. Theory and experiment agreed to an accuracy of one part in ten decimal places, or 0.00000001 per cent. This is the most precise agreement between theory and observation for any experiment carried out. This is powerful evidence that the whole edifice of quantum physics is built on solid foundations, and we can use it with confidence to predict (or explain) the behaviour of atoms and molecules.

splitting hairs
In order to equal the accuracy of calculations carried out using QED, the distance between New York and Los Angeles, which is 3961 km (2462 miles), would have to be measured to within the thickness of a human hair, which is 0.2 mm ($^1/_{127}$ inch).

putting quanta to work

One result of quantum uncertainty (the principle that says that a particle cannot have position and momentum simultaneously) is a process called tunnelling. This explains among other things how the Sun shines. Stars, such as the Sun, release energy by the process of nuclear fusion. In order for nuclei to fuse, two positively charged nuclei (in the simplest case, two nuclei of hydrogen each consisting of one proton) must come together. But according to classical electromagnetic theory this is impossible because the two positively charged particles repel one another and cannot fuse. Quantum physics provided an explanation of how fusion might be possible. Due to quantum uncertainty, if two protons are very close together it is not certain if they are touching or not, and they may or may not fuse. Another way to think about this is in terms of waves. If the quantum entities can approach each other so closely that their wave functions overlap, they can be pulled together by the interacting wave functions. This is called "tunnelling" because the two particles come up against the electric barrier of classical physics that exists between

.... helium nucleus (two protons and two neutrons)

.... photon is emitted

.... sunlight

quantum explanation
A star such as the Sun releases energy by forcing protons, four at a time, to combine to make helium nuclei. Until quantum theory was applied to the problem, this was a great puzzle, because the temperature inside the Sun is not high enough to do this without the help of quantum uncertainty.

them, but they are able to "tunnel" through it.

If quantum physics did not operate inside stars, the Sun would not shine and we would not be here. The same tunnelling process operates in reverse, enabling particles to get out of nuclei in the process of radioactive decay, or in nuclear fission. Particles in a nucleus are held together by a force called the strong nuclear force, which has a very short range (so nuclei are small) but overwhelms the electric repulsion of protons and neutrons once they are touching one another (just as you can overwhelm the expansionist tendency of a spring by squeezing it with your fingers). The effect is as if the particles are sitting in the deep crater of a volcano, which they would have to climb to escape, but once over the lip of the volcano they would be rapidly repelled by the electric force. Tunnelling enables some particles to escape from some nuclei without having enough energy to climb this barrier.

under pressure
The springiness of a spring can be overcome by a stronger squeeze, just as the electric repulsion between protons in a nucleus can be overcome by the strong force.

random decay

But there's a further oddity about this process. In a collection of potentially radioactive nuclei, they do not all decay in this way at once. Individual nuclei decay at random while obeying the precise laws of chance. For a particular radioactive element, there is a characteristic

tunnelling out
Particles trapped inside the nucleus of an atom by the strong force can escape because of the waviness associated with quantum uncertainty.

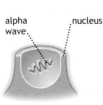

alpha wave.   .nucleus

an alpha wave has insufficient energy to escape the strong force of the nucleus

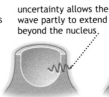

uncertainty allows the wave partly to extend beyond the nucleus.

there is a measurable probability that part of the wave extends outside the nucleus

wave escapes.

the wave is then able to "tunnel" out and escape from the nucleus completely

time called the half-life. In one half-life, half the nuclei decay. In the next half-life, half the rest decay, and so on. So if you start with 128 nuclei with a half-life of 10 minutes, after 10 minutes 64 have decayed, in the next 10 minutes 32 decay, in the next 10 minutes 16 decay and so on (these are all average figures; in a real sample the numbers decaying may differ slightly). But it is impossible to say in advance whether any particular nucleus will decay immediately, in 10 minutes' time, in half an hour, or whenever. Decay is as random as tossing a die, and Einstein hated the idea. So did Schrödinger, who picked up on this and combined it with the idea of the collapse of the wave function in his famous Cat Paradox (see page 32).

chain reaction
In a nuclear reactor, unstable nuclei of radioactive elements (such as uranium) absorb neutrons, making them "split" into two almost equal parts, releasing energy. As by-products of this fission, more neutrons are released, which in turn encourage more nuclei to split. A balance is maintained when each nucleus that splits triggers the splitting of just one more nucleus – thus creating a sustained chain reaction.

the hydrogen bond

Quantum physics also operates at the heart of life. The way atoms combine to form molecules is fully explained by quantum physics, which describes how electrons are shared between atoms to achieve the most stable combinations. The links between atoms that form molecules are called bonds, and they involve only the outermost electrons in the cloud around a nucleus. But there is one special case. A hydrogen atom consists of a single proton and a single electron, which, thanks to its quantum mechanical nature, still manages to surround the proton. There are no inner layers of electrons to conceal the proton when its lone electron is being shared with another nucleus in a conventional chemical bond.

decay and half-life

When an individual atom (strictly speaking, a nucleus) undergoes radioactive decay, it does so in accordance with the rules of chance. It is as if the atom tossed a coin, and decayed if it came down heads, and didn't decay if it came down tails; or as if it rolled a die, and only decayed if an even number comes up. Then it makes another toss, or roll. This means that in a large number of atoms, a certain proportion will always decay in a certain time. Physicists measure this in terms of the time it takes for half the atoms in a sample to decay, and they call this the half-life. Each type of radioactive nucleus has its own distinctive half-life.

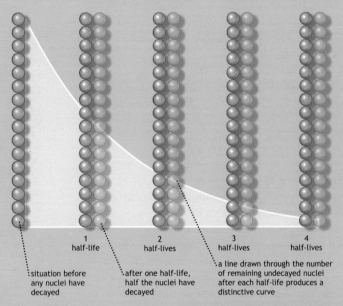

1 half-life

2 half-lives

3 half-lives

4 half-lives

situation before any nuclei have decayed

after one half-life, half the nuclei have decayed

a line drawn through the number of remaining undecayed nuclei after each half-life produces a distinctive curve

radioactive decay

The length of time covered by a half-life can range from less than a millionth of a second to millions of years depending on the element. The shorter the half-life, the more radioactive the element is. For example, Nitrogen-16, which occurs in nuclear power stations has a half-life of seven seconds while Carbon-14, which is used in radio-carbon dating, has a half-life of 5730 years.

The result is that the positive charge on the proton is partly exposed to the outside world – but only partly exposed, because the wave function of the electron still extends around the proton, even though it is more strongly concentrated in the bond. In the classical view of electrons, this could not happen; the electron would have to

...in liquid water, molecules have looser bonds and evaporate from the surface

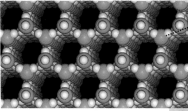

...when frozen, water molecules crystallize into an open lattice

be on one side of the nucleus or the other. As a result of the partial shielding, hydrogen atoms that are already involved in chemical bonding on one side effectively show a fractional positive charge on the other side, and this can attract the electrons in a cloud around another

ʾʾ When it comes to atoms, language can be used only as in poetry. The poet, too, is not nearly so concerned with describing facts as with creating images. ʾʾ

Niels Bohr (1922)

molecule. The result is that they can form a so-called hydrogen bond, a kind of bridge between the two molecules, which is weaker than a conventional bond.

The strength of the hydrogen bond depends on the quantum properties of electrons and can

be calculated; it is about one-tenth as strong as a conventional bond, and this matches experimental measurements, again confirming that quantum physics works in the real world. But the key importance of the hydrogen bond is that it is what holds deoxyribonucleic acid (DNA) together. DNA, the molecule of life, is made of two intertwined strands (the famous double helix) held

liquid and solid
In liquid water, permanent hydrogen bonds cannot form between the molecules because they possess too much energy and move too quickly. When water freezes, however, permanent hydrogen bonds lock into place, crystallizing the molecules in a very open lattice. This is why ice at 0°C is less dense than water, and why ice floats on water.

single strand
of DNA

"ladder" twists
up to make
double helix

hydrogen bonds
link sub-units to
make "rungs"

bonded rungs

The famous double helix of the DNA molecule is like a twisted ladder. The rungs of the ladder are held in place by hydrogen bonds, which link the half-rung from one strand of the molecule to the half-rung on the other strand.

hydrogen bonds
allow the strands
to separate

together by hydrogen bonds running like a zipper down the middle of the helix. When genes are being read off the DNA and used by the machinery of the cell, it is the relatively weak hydrogen bonds that are unzipped in the process and zipped up again when the job is done. In certain forms of genetic engineering, the machinery of the cell is used to cut out pieces of unzipped DNA completely, and replace them with other genetic material. It is the quantum nature of electrons that makes both life itself, and genetic engineering, possible.

laser technology

The quantum properties of electrons are also important in electronics, and especially in the design of microchips. The behaviour of some circuits, for example, depends crucially on the ability of electrons to tunnel through barriers. But one of the neatest examples of the application of quantum physics in everyday life is the laser, a device found in every CD player.

When a collection of atoms or molecules gets hot, they pick up energy (they are said to be "excited"), and the electrons in them are raised up to higher energy levels. Left to cool off, these

high energy

An electron jumps up to a higher energy level in an atom when it is excited, in the same way as a child jumps for joy.

electrons jump back down to lower energy levels radiating light in a more or less random fashion to produce the black body curve (see page 14). But if a weak input of radiation with just the right energy is fed into the right kind of

laser display
The first public performance of a laser show took place on May 9, 1969, at Mills College in Oakland, California. Today, the laser beams are made to move by mirrors controlled by magnets and electronics that can move the beam faster than the eye can follow.

material, the effect is to raise the electrons in each atom (or at least, very many atoms) of the material up into the same excited state as all their neighbours. Now, when they jump back down from the excited state they will each emit a photon with exactly the same amount of energy – which means precisely the same wavelength. It is all these billions of photons marching precisely in step with one another that form the intense beam of pure monochromatic light of a laser (the name comes from Light Amplification by Stimulated Emission of Radiation).

The principle is very simple – it actually goes back to some calculations made by Einstein in 1916. But the technological difficulty of getting a lot of atoms into the same excited state and keeping them there until you are ready to trigger the release of energy proved so great that the first lasers were not developed until 40 years after Einstein's calculations. Now, you can get one given away free with a piece of hi-fi, to hang on your key-ring. There is no better example of how far quantum physics has become part of everyday life since the days of Einstein and Bohr. But there are still aspects of the quantum world to be explored and turned into new technologies for the future. We discuss some of these ideas in the next section.

miniature version
A tiny pocket laser attached to a key-ring is an example of quantum technology, based on an idea from Einstein that is now commonplace.

how lasers work

Laser light is made by exciting electrons with a weak energy source. When many electrons are excited, another pulse of energy acts as a trigger so they all fall to the same lower level at the same time. This means they all emit the same wavelength of light in a concentrated pure beam.

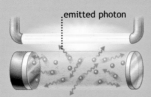

non-lasing state
The laser apparatus consists of a ruby rod with mirrors at each end and a light source next to it.

Labels: weak light source, mirror, half-silvered mirror, ruby rod

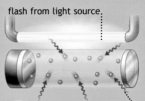

switched on
When the flash tube fires, light enters the ruby rod and "excites" electrons in many of the atoms.

Labels: flash from light source, atoms absorb energy

emitting light
Some of the excited electrons fall back to lower energy levels, emitting photons in all directions.

Label: emitted photon

parallel photons
Photons travelling along the rod bounce between the mirrors, triggering more atoms to emit photons.

laser surgery
Lasers have many uses in eye surgery, which include re-attaching detached retinas and cauterizing blood vessels in the retina.

laser beam
When the energy has built up enough, it is released from one end as a beam of intense single-wavelength laser light.

Label: laser light

inside the nucleus

By the 1960s, there was indirect evidence from experiments that the protons and neutrons that make up the nucleus of an atom are not truly fundamental (indivisible) particles, but possess an inner structure. By contrast, the electron does seem to be a fundamental particle, with no internal structure. Over the years, several ideas were suggested as models for what goes on inside protons and neutrons, which collectively are known as nucleons. One of these possibilities, which became known as the quark model, was first invented by George Zweig in 1963. Zweig worked at Caltech and then at CERN in Geneva. Another model was developed by Murray Gell-Mann, who worked independently at Caltech and was unaware of the work of Zweig, who had left before publishing his results. At the time, this was only one candidate among several speculative models, but is singled out because, years later, experiments confirmed it to be a good description of what goes on inside protons and neutrons. These experiments were carried out at the Stanford Linear Accelerator (SLAC) in California at the end of the 1960s. They involved shooting electrons down an evacuated (low pressure) tube 3.5 km (roughly 2 miles)

finding quarks
The Stanford Linear Accelerator (SLAC) is housed in a tube two miles long, making a straight line across the California countryside (top photograph). The lower picture shows scientists working in the tunnel of the accelerator. This machine provided the first hard evidence for the existence of quarks.

in length, and observing how they scattered from nucleons – protons or neutrons. The data showed that the electrons were bouncing off little hard particles within the nucleus. Richard Feynman called these particles "partons", a name chosen to avoid favouring any of the speculative models, and simply meaning that the particles are "part" of the nucleon. But it soon became clear that the behaviour of these partons exactly matched the predictions of quark theory.

large detector
A giant solenoid magnet lies at the heart of the SLAC Large Detector; this device monitors the particles produced in the Linear Accelerator.

colours and glue

In its modern form, the quark model says that each nucleon is composed of three quarks (taken from "Three quarks for Muster Mark" in James Joyce's *Finnegans Wake*). The quarks have a property analogous to electric charge, but occurs in three varieties, not two (positive and negative). These varieties are labelled using the names of colours – purely for convenience. Quarks are not really coloured, and the labels could have been Tom, Dick, and Harry. The names of the colour charges are actually red, blue, and green. There is always one quark of each colour in a nucleon, and the three colours in effect cancel each other out so that they have only a slight influence on other particles. However, the quarks can have electric charge as well. Just as the electromagnetic forces

ups and downs
A neutron has two "down" quarks with a charge of $-1/3$, and an "up" quark, with a charge of $+2/3$, giving it no charge. A proton has two "up" quarks and one "down" quark, giving an overall charge of +1. "Up" and "down" are whimsical names given by physicists.

down quark   up quark

$-1/3$ $+2/3$

$+2/3$ $-1/3$

$-1/3$ $+2/3$

neutron: charge = 0   proton: charge =+1

asymptotic freedom

Quarks inside a nucleon (neutron or proton) are confined as if they were held together by lengths of immensely strong elastic. When the quarks are close together, the elastic is floppy and they move freely. This is called asymptotic freedom. If you try to pull them apart the elastic stretches, and the harder you pull the more strongly they are pulled back, which is why a single quark cannot exist on its own.

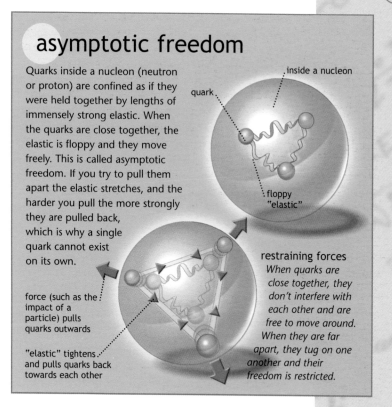

inside a nucleon

quark

floppy "elastic"

force (such as the impact of a particle) pulls quarks outwards

"elastic" tightens and pulls quarks back towards each other

restraining forces
When quarks are close together, they don't interfere with each other and are free to move around. When they are far apart, they tug on one another and their freedom is restricted.

between charged particles are carried by photons, so the colour forces between quarks are carried by particles analogous to photons, whimsically called "gluons" because they hold quarks together, and this glue is very strong. Whereas the electric force gets weaker when charged particles are further apart, the glue force gets stronger as the two quarks are pulled further apart in the same way that it becomes harder to stretch a piece of elastic the more it's stretched. The result is that no quark can ever escape from the nucleon to have an independent existence of its own.

quark

nucleon

glue force
leaks out

**sticking
together**
*The glue force
that holds quarks
together inside a
nucleon can leak
out of that nucleon
and affect quarks in
the nucleon next
door. It is this
leakage of the glue
force that we detect
as the strong force
that holds atomic
nuclei together.*

**tracking down
the boson**
*The UK physicist,
Peter Higgs,
developed a model
to explain why
particles have mass,
this predicts
the existence of a
kind of particle –
now called the
Higgs boson –
but scientists
are still
searching
for the
Higgs
particle.*

This whole package of ideas is closely modelled on
quantum electrodynamics, and is known as quantum
chromodynamics, or QCD. The glue force between
quarks leaks out from one nucleon to affect its
neighbours relatively slightly, and provides the
strong force which holds the nucleus
together. The real colour force is a lot
stronger than the "strong" nuclear force,
which overwhelms the electrical repulsion
of protons on the nuclear scale.

nuclear decay

There is just one other interaction that affects the
behaviour of particles in the quantum world, and the
way it works highlights why physicists prefer the term
"interaction" to the term "force". The weak interaction,
as it is known, also operates only on the scale of nuclei
and nucleons. Its most important property is that it is
responsible for some of the processes involved in
radioactive decay and this can best be seen from an
example – the process known as beta decay, in which
a neutron is transformed into a proton.

The process derives its name from when this kind of
radioactivity was first discovered in the closing years of
the 19th century. Then, the radiation produced was called
beta radiation (another kind of radiation had already been
dubbed alpha radiation); it was
only later that it was
discovered that "beta
rays" are in fact a
stream of fast-
moving electrons.
In beta decay,
a neutron (either
sitting on its
own in space, or

inside a nucleus)
transforms itself
into a proton by
spitting out an
electron and a
particle known as
a neutrino. It is
important to
appreciate that there is no

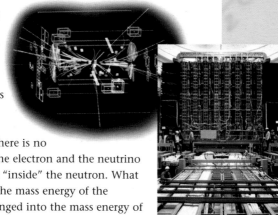

sense in which the electron and the neutrino
were at any time "inside" the neutron. What
happens is that the mass energy of the
neutron is rearranged into the mass energy of
a proton plus the mass energy of an electron
and a neutrino. The way in which the neutron does this
is by ejecting (from one of its constituent quarks) the
particle that is the "force carrier" of the weak interaction,
analogous to the photon in QED and the gluon in QCD.
But this particle, which is rather boringly called an
intermediate vector boson, carries one unit of negative
electric charge. The energy of the boson is then rapidly
converted into the mass energy of a neutrino and an
electron. The electron has the unit of negative charge, so
to balance the books the neutron left behind now has one
unit of positive charge and has become a proton. There is
also a similar boson, which carries one unit of positive
charge, and one which, like the photon, has no charge at
all; these are involved in other kinds of weak interaction.

The discovery of all three of these bosons in
experiments at CERN in 1983, following the triumph of
QCD in the 1970s, set the seal on what has become the
standard model of the quantum world. Everything on
Earth can be explained in terms of quarks, electrons and
their associated neutrinos, and four interactions – gravity,
electromagnetism, the weak nuclear interaction, and the
glue force underpinning the strong nuclear interaction.

picturing particles
*The UA1 detector
at CERN, (pictured
above during its
construction in
1981), is the size
of a house. When
beams of subatomic
particles collide in
the centre of the
detector, they
produce showers
of other particles,
which are tracked
by the UA1
assembly. A
computer then
produces a picture of
the particle shower.
The image (above
left) records the first
detection of a Z
particle, one of the
intermediate vector
bosons, in 1983.*

the quantum computer

A prediction was made in 1965 by Gordon Moore, the co-founder of Intel, who said that the number of transistors in computer chips would double every 18 months. This prediction, now known as "Moore's Law", has so far proved to be correct, but the law will soon expire, as a transistor is now being developed with components only three atoms thick. Before long, the physical limitations of using atoms to make working parts will be reached. The next logical step is to use atoms and molecules themselves to perform memory and processing tasks inside a quantum computer. Whereas computers now manipulate "bits" that exist in one of two states – 0 or 1 – quantum computers encode information as quantum bits, or qubits. A qubit can be a 1 or a 0, or in a superposition that is simultaneously both 1 and 0, or somewhere in between. As a quantum computer can contain these multiple states simultaneously, it has the potential to be millions of times more powerful than today's most powerful supercomputers.

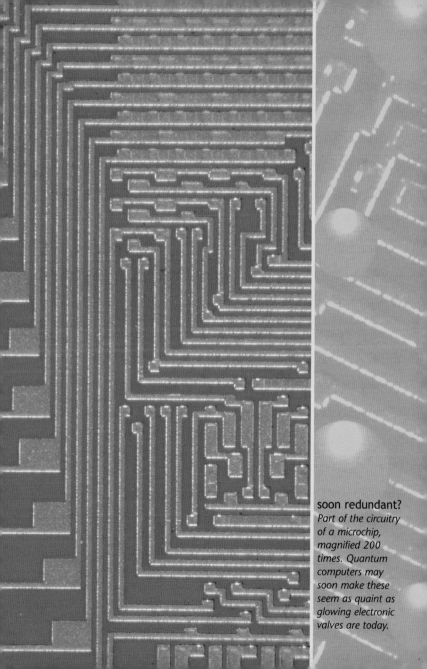

soon redundant?
*Part of the circuitry
of a microchip,
magnified 200
times. Quantum
computers may
soon make these
seem as quaint as
glowing electronic
valves are today.*

principles into practice

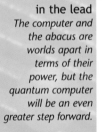

Modern electronic computers already depend on the quantum properties of electrons and atoms for the way their microchips work. But teams around the world are already working on the possibility of a true quantum computer, which would be as much a development from today's best computers as today's computers are a development from the abacus. The best way to picture how such a device would work is in terms of the idea of a superposition of states. Physicists still argue about whether an entity such as a cat can be in a superposition of states, but there is no doubt that a single electron can exist in this simultaneous duality. For example, an electron in an atom might be able to exist in either of two states – the ground state of lowest energy, or an excited state that it can jump to if it has the energy. These states could correspond to the numbers 0 and 1 of the standard binary code of computers. Such a single

in the lead
The computer and the abacus are worlds apart in terms of their power, but the quantum computer will be an even greater step forward.

the near future
A vial containing quantum computer molecules is loaded into a nuclear magnetic resonance apparatus, where a program of radio-frequency pulses will direct the molecules to carry out a calculation.

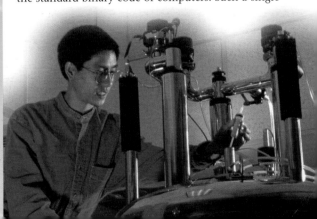

quantum unit of information is sometimes called a quantum dot. If light of exactly the right wavelength is shone on the atom for precisely the correct length of time, it is possible to create a situation where there is a 50:50 chance of the electron being in either of these states. In quantum physics, this is the same as saying that the wave function of the electron is a 50:50 mixture of both states – a superposition of states.

ever smaller
The chip being carried by this ant is more powerful than a 1960s room-sized computer. Quantum systems, however, will allow far greater miniaturization when one electron will carry out the work of millions.

current speed limits

The fact that quantum dots can exist in a superposition of states opens up the possibility of making powerful computers by using a small number of physical components. What matters in determining the power of a computer is the number of on/off switches that make up its memory and determines how long the strings of 0s and 1s can be in the binary code of its programs, its calculations, and its answers. Even with microchip technology, there are limitations on the power of computers because electrons have to travel from one switch to another inside the memory. This takes time, which limits the speed of the computer. The flowing electric currents also generate heat, which limits how large the memory can be in physical terms without melting. Computer scientists are trying to get round these problems by using devices such as laser beams to carry the information, but even this kind of technology is limited compared to the power of quantum computing.

too hot to handle
Miniaturization brings its own problems. With components packed closely together, one of the limits on modern PCs is how hot they get.

exponential growth

Imagine the possible choices available to a "computer" consisting of only one pair of quantum dots. Each dot can be set at either 0 or 1. So there are four possible states the computer can be in – 00, 01, 10, and 11. A conventional computer can only be in one of these states at any one time. However, in a quantum computer where both quantum dots are in a superposition of states, it is in effect operating in all four states simultaneously, as if it were four conventional computers wired together. In general, the number of states simultaneously available to the quantum computer is 2 (because it is using binary language) raised to the power of the number of quantum dots (in this case, also 2). With just three dots, the computer would explore 8 states at once (2^3), and so on. The figures in such an exponential law very quickly take off and mean that a quantum computer made up of just 10 quantum dots joined together would be as powerful as a conventional computer with 1,024 switches in its memory – exactly one kilobit. A quantum computer with one kilobit of physical memory would act like a

quantum quadruple
Four computers for the price of one – in a quantum computer, each quantum binary digit (qubit) is equivalent to four conventional on/off switches (bits)...

geometric increase
... but if you have twice as many bits, you don't double the power of the quantum computer – you square it. The power grows exponentially, so just 10 quantum dots (qubits) are equivalent to a current kilobyte.

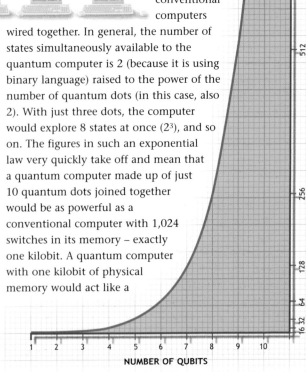

NUMBER OF QUBITS

virtual machine with 10^{1024} bits of memory, a number vastly greater than the number of atoms in the Universe. It would be more powerful than a conventional computer made up of every atom in the visible Universe.

code-breaking

There are problems in learning how to connect the quantum dots in the right way, how to program such machines, and how to interpret their output. But solutions are being found with quantum computers made up of only a few dots, and physicists know that the effort is worthwhile because it has been proved mathematically that full-size quantum computers will work. A key feature of this highly technical proof came from the work of Peter Shor at the AT&T Bell Laboratories in New York.

When **Peter Shor** (born 1959) looked into the problem of cracking codes that involve very large prime numbers, he was able to show that a quantum computer could solve in seconds a problem that would take a conventional computer months. Now, all he has to do is build a quantum computer!

Shor was interested in how quantum computers might be used to crack codes. There is a particular type of code that is based on multiplying very large prime numbers together. The result is a much larger number, and the code can only be cracked by finding the two prime numbers that produced it – a process that is called factorizing. This process takes months, or even years using conventional computers. In 1994 Shor proved that a quantum computer could solve this problem

" The fact that such extremely simple experiments can lead to such tremendous conclusions about the nature of reality, and the depth of those conclusions, is really the strangest thing I know. It's strange, and it's weird. But not, as many people say, mysterious – because I think we understand it. "

David Deutsch, Oxford physicist, (2000)

by "guessing" possible factors and multiplying them out to see if they give the extremely large number. This is more or less the way a conventional computer works, checking one possibility after another, but in a quantum computer each guess (a string of 0s and 1s) would correspond to one of the superposed states, so the computer would make all the guesses and carry out all the calculations simultaneously. Shor showed that all the incorrect guesses would cancel each other out (like the destructive interference that makes dark stripes in the double-slit experiment), while the right answers would reinforce one another (like the constructive interference that makes bright stripes in the double-slit experiment), so that just one solution would emerge from the machine. All the wrong answers destroy one another.

permissible errors

A perfect quantum computer would be even more awesome. It would simultaneously solve every problem that it was capable of solving, and give the answer to every question that it was capable of answering, providing we knew how to program it and how to interpret the output. Such a prospect is a long way off, but in a sense it means that you need only ever build one full-size quantum computer and run it once, then spend generations trying to understand the output.

the answer is **42**!

But at the level we are likely to achieve in our own lifetimes, it wouldn't matter if the machine makes mistakes. A few of the quantum dots might, for example, fall out of their superposition of states due to the effects of random noise inside the machine and lead to wrong answers being generated. Such a machine would be so fast, though, that it would only need to produce the right answer once in a thousand times. To overcome this particular problem, you could run the computer a thousand times before breakfast and then multiply all the possible answers on an ordinary computer to see which factors really do produce the extremely large number that you are interested in.

"Classical information is like the information in a book... the information in real microscopic physical systems is more like the information in a dream. It's certainly there in some sense, but if you try to tell somebody about your dream you won't remember it the same way it was before you told them."

Charles Bennett, US physicist (2000)

Such an imperfect but very fast computer would be immensely valuable in the areas of military, political, and industrial espionage, and there is no doubt that the existing funding for such projects will continue. In this area of technological development, the less that is heard about it, the more likely it is that progress is being made. And once imperfect quantum computers exist, their reliability will improve inexorably. Then, their immense power might be able to help us turn other weird quantum possibilities into practical reality.

old style espionage
Applying the power of a quantum computer to the discovery and decoding of secret information could render the role of the traditional spy obsolete.

teleportation, quantum style

The strangest thing about the quantum world is the way an entity such as an electron seems to be in several places at once. In the double-slit experiment, for example, a single electron passing through the apparatus seems to "know" about the entire experimental set-up, and its own place within it. This is called non-locality, because the electron is not localized at any single point. It is only when the particle interacts with something (such as the TV screen) that the wave function collapses and it becomes localized.

same time, different place
In the quantum world, common-sense ideas about "locality" break down and a particle really can be in more than one place at one time.

the phenomenon of non-locality

Most electrons are interacting with something all the time. An electron in an atom is interacting with the nucleus of the atom, so it is localized not only to stay in the vicinity of the atom, but at a specific energy level as well. But quantum entities can be prepared in states where they are interacting only in a very limited way with other quantum entities, and then non-locality shows up with full force.

Experiments to demonstrate quantum non-locality were devised by the Irish physicist, John Bell, working at CERN in Geneva in the mid-1960s. Several experimental teams took up the challenge of turning these ideas into reality, and the definitive demonstration was carried out by a group working in Paris in the early 1980s. In these

collapsing colours

Quantum non-locality can be described metaphorically in terms of coloured balls. Imagine that balls come in only two colours, blue or yellow. Because of the "law of equal colour" an atom can only eject two balls at a time, and there must be one of each colour.

But quantum physics says that the colour of a ball is not decided until it is measured. The two balls flying away from the atom in opposite directions exist in a superposition of states, blue and yellow mixed together to make green, until the measurement is made. Now suppose you measure the colour of one of the balls. The wave function collapses, and it is seen to be, say, blue. But by the law of equal colour, that means that the other ball, far away, must collapse into a state of yellowness, at exactly the same time. The balls behave like one particle even though they are not in the same location.

atom

atom throws two balls outwards....

emitting quantum entities
Two "balls" with undetermined colour are ejected from an atom.

flying apart
No matter how far they fly, neither ball "knows" what its own colour is.

each ball is in a mixture of "blueness" and "yellowness" unless it is observed

long distance effect
At the instant one ball is seen to be yellow, the other becomes blue, without anyone looking at it.

the moment it is observed, the ball collapses into one of the two possible colours

Irish physicist **John Bell** (1928–90) devised an experiment to prove the non-local nature of the world. His work showed that it would be possible, in principle, to make measurements on two quantum entities (such as photons) that had once been in contact, and discover if they remained "aware" of each other when far apart. Such experiments were actually carried out by Alain Aspect in Paris, and proved conclusively that quantum systems are non-local.

experiments, an atom is induced to emit two photons simultaneously in opposite directions. The common origin of these photons means that they are correlated with one another and, according to the equations of quantum physics, they remain "entangled" even when they are far apart – as if they formed a single particle. The experiments proved that measuring the properties of one of the photons on one side of the lab affected the other photon on the other side of the lab instantaneously. Through these experiments non-locality could be seen at work.

By the middle of the 1990s, researchers in Geneva had extended such experiments so that they involved sending photons along fibre-optic cable 10 km (6.2 miles) in length. The experiments still show non-locality. The two photons behave like one particle even when they are 10 km apart. And it cannot be over-emphasized that these are genuine experimental results, not "merely" the predictions of the theory.

conveying coded messages

But although some influence links the two photons instantaneously, no useful information travels between them faster than the speed of light. The measurement of photon A disturbs photon B in a random fashion. Observations can show that photon B has been disturbed, indicating that something happened to photon A, but cannot reveal precisely what it was that happened to photon A. If, however, some (incomplete) information about what was done to photon A is sent to us by a conventional route (by letter, e-mail, or carrier pigeon), we can put this together with our observations of photon B to end up with more information than we would get by one of the two routes alone. This has implications for cryptography and is attracting considerable funding as a result. In principle, a secret message can be sent in two parts, neither of which makes sense on its own, but one of which involves quantum entanglement. The whole message travels no faster than light because both halves are needed, but the quantum entanglement cannot be intercepted without changing it.

It is even possible for the "message" to be the original particle, photon A, itself – a form of teleportation, although not quite the way it happens in *Star Trek*, leading many physicists to quip, "It's teleportation, Jim, but not as we know it." The possibility was pointed out by Charles Bennett of the IBM Research Center at Yorktown Heights, New York, in a paper published in 1993. Since then, experiments have confirmed that the idea works, at least on the laboratory scale.

half a message can travel by conventional means...

while the other half goes faster than light

a message in two halves
Quantum non-locality means that when one "entity" is tweaked, another instantaneously responds, no matter how far away it is. To understand that response – and its meaning – you have to wait for further information to arrive by a conventional route.

more than a copy

The idea depends on the fact that if one entity is indistinguishable from another entity in every way, then it *is* that entity. When you make a photocopy, you have two "entities" and you know which is the original. But if the original were destroyed in the process, and the "copy" was identical in every way to the original, then it would not really be a copy at all. It would *be* the original.

Quantum teleportation works like this. Two entangled photons are prepared in the usual way. One is carried off to a distant place (the Moon, perhaps). The other one is

allowed to interact with an electron, and all the information about this interaction (which destroys the original quantum state of the photon) is stored. The interaction will have changed the quantum state of the photon in a box on the Moon, but nobody on the Moon knows this yet. Now, the results of the interaction between the electron and the photon back on Earth can be sent to the Moon, by rocket, or laser beam, or any conventional method that does not involve travelling faster than light. Armed with this information, a skilled physicist on the Moon could tweak the photon in the box in such a way that the tweaking subtracts out the changes caused by the entanglement, and produces an exact copy of the first photon. Indeed, by any conceivable test it is the first photon.

It should be emphasized that this really has been done, for pairs of photons separated by a few metres. At one level, this is a rather futile thing to do for a photon, because the whole process takes place much slower than light speed, and it would be quicker to send the photon across the room in the conventional manner. But it demonstrates that in principle it is possible for an exact copy of a physical

American-born physicist **David Bohm** (1917–92) was one of the first people to appreciate the essentially non-local nature of quantum physics. His ideas were largely ignored from the 1950s to the 1980s, but were rehabilitated thanks to John Bell and Alain Aspect. They are at the forefront of 21st century thinking about practical applications of quantum phenomena.

object to be transported in this way any distance across space, with the proviso that the whole process must take place at less than light speed.

taxing quantum matter

Just how soon any of these ideas become practicable in the world outside the laboratory remains to be seen, but there can be little doubt that they will affect our lives in the not-too-distant future. More than 150 years ago, not long after he had invented the electric motor, Michael Faraday was asked by a leading politician of the day what use his invention was. He replied to the effect that he had no idea what practical use his invention might be put to, but that he was sure politicians would find a way to tax it.

> **I think I can safely say that nobody understands quantum mechanics… do not keep saying to yourself, if you can possibly avoid , 'but how can it be like that?' because you will go 'down the drain' into a blind alley from which nobody has yet escaped. Nobody knows how it can be like that.**
>
> Richard Feynman, *The Character of Physical Law*, BBC (1965)

It surely won't be too long before taxes are being paid on the income generated by quantum computers, quantum cryptography, quantum teleportation, and other as yet undreamed-of devices. Almost instantaneous travel, genuinely intelligent computers, and perfect "virtual realities" are probably just around the corner. As Arthur C. Clarke said, "any sufficiently advanced technology is indistinguishable from magic." We will have magic at our fingertips within a hundred years.

an unimaginable future
The electric motor started out as a toy. Even Michael Faraday, who invented it, could not predict that it would lead to trains that travelled at 350kph (217mph).

new horizons
Before the invention of the steam engine, the human world had scarcely changed for thousands of years. Now, the power of the quantum is opening the door to an era of unprecedented change and unimaginable possibilities.

glossary

alpha particle
Entity made up of two neutrons and two protons bound together by the strong nuclear force. Equivalent to a helium nucleus (a helium atom without its electrons), but such a stable unit that it behaves like a particle in its own right.

asymptotic freedom
Technical term for the way the glue (or colour) force between quarks decreases the closer they get together.

atom
The smallest building block of a chemical element. All atoms of any particular element are identical to one another. Each atom is made up of a tiny central nucleus, carrying positive electric charge, surrounded by a cloud of negatively charged electrons. The number of electrons in the cloud balances the amount of positive charge in the nucleus, so the atom is electrically neutral overall.

beta radiation
Old name for a stream of fast-moving electrons.

black body
An object that is a perfect absorber of radiation.

black body radiation
Radiation emitted by a hot black body.

Bohr radius
The minimum distance at which an electron can "orbit" an atomic nucleus, in the model devised by the Dane Niels Bohr. Although the model is an imperfect description of the atom, this number – 5.29×10^{-11}m – is still a good indication of the size of an atom.

cavity radiation
Old name for black body radiation.

classical mechanics
The description of the workings of the everyday world in terms of Newton's Laws and Maxwell's Equations.

classical physics
The laws of physics that apply to smoothly changing phenomena, especially Newton's Laws and Maxwell's Equations.

colour
A property of quarks equivalent to the charge on electrically charged particles. Also known as colour charge. It has nothing to do with colour in everyday life.

Copenhagen Interpretation
The package of ideas describing the quantum world in terms of probability, uncertainty, and the "collapse of the wave function".

diffraction
The way waves bend round corners, or spread out from a small hole, as in the double-slit experiment.

electron
A fundamental entity, identified in the 1890s by the British physicist J. J. Thomson as a piece of matter that is a component of the atom. Usually thought of as a tiny particle, but one of the startling discoveries of quantum physics is that electrons also behave as waves.

element
A substance that cannot be broken down into a simpler substance by chemical means.

energy level
A quantum state with a particular energy associated with it; most commonly used to describe the quantum states available to an electron in an atom. An entity such as an electron jumps from one level to another without passing through any intermediate state; this is the famous quantum leap.

Feynman diagram
A representation of the way particles interact with one another by the exchange of force-carrying quanta, such as photons.

field

The extended influence of a force, such as electro-magnetism, through space.

gamma ray

Electromagnetic radiation with very high energy, corresponding to very short wavelengths, from 10^{-10} to 10^{-14} of a metre.

gluon

Quantum entity that plays the equivalent role in quantum chromodynamics that the photon plays in quantum electrodynamics.

h

Letter used to denote Planck's constant.

half-life

The time it takes for half of the atoms (strictly speaking, nuclei) in a sample of radioactive material to decay.

interaction

Any of the four forces of nature – gravity, electro-magnetism, and the strong and weak nuclear forces.

interference pattern

Pattern produced by interference, as in the double-slit experiment.

intermediate vector boson

Overall name for the three particles (W^+, W^-, and Z^0) that are the carriers of the weak interaction, carrying out the equivalent to the role of photons in the electromagnetic interaction.

jump

See quantum leap.

laser

Powerful beam of light of a single wavelength. (Light Amplification by Stimulated Emission of Radiation.)

light quantum

Photon.

matrix mechanics

A version of quantum mechanics based on equations, formulated in terms of entities known as matrices.

momentum

The force that keeps an object in motion.

neutron

One of the particles that make up the nucleus of an atom. Roughly the same mass as a proton, but with no electric charge.

Newton's laws

The laws of mechanics that apply in the everyday world, describing such things as how billiard balls bounce off one another, and how rockets work.

non-locality

The way a quantum entity, such as an electron, "spreads out" between interactions, so that it cannot be said to be localized at a point.

nucleon

Generic name for protons and neutrons, which together make up the nucleus of an atom.

nucleus

The central core of an atom, made up of positively charged protons and electrically neutral neutrons.

orbit

In general, the trajectory of any object moving under the influence of another object. In quantum physics, usually referring to the trajectory followed by an electron under the influence of the nucleus of an atom.

photoelectric effect

Ejection of electrons from a metal surface when light shines on the surface.

photon

Particle of light. Also the carrier of the electro-magnetic force.

Planck's constant

Fundamental constant that relates the energy of a photon to its frequency, through the equation $E = hf$. The constant, h, appears in many equations of quantum physics.

qubit

Quantum equivalent of a bit (binary digit) in computing. A bit can only have the value 0 or 1. A qubit can also exist in a mixed state, partly 1 and partly 0.

quantum

The smallest unit of something that it is possible to have. Originally used for the quantum of light, now called a photon.

quantum chromodynamics (QCD)

Theory of how quarks interact by the exchange of gluons. Name chosen by analogy with quantum electrodynamics.

quantum dot
The equivalent of a binary on/off switch in a quantum computer. Unlike switches in the everyday world, a quantum dot can exist in a superposition of states.

quantum electrodynamics (QED)
The quantum theory of how charged particles and photons interact with one another.

quantum leap
The smallest change it is possible for a system to make. Such change involves a random choice from a set of possible changes.

quantum mechanics
The description of the workings of the world on very small scales in terms of the laws of quantum physics.

quantum physics
The laws of physics that apply on very small scales (typically, the scale of atoms and below) where change does not occur smoothly but in discrete steps, or quanta.

quark
Fundamental particle at a level below that of the proton and the neutron. There are three quarks in every proton, and three quarks in every neutron.

radioactive decay
The process whereby an unstable particle or atomic nucleus transforms itself into a different particle (or nucleus), usually by spitting out another particle,

such as an electron or an alpha particle.

radiation
Any kind of energy travelling through space. Light is a form of electromagnetic radiation.

Schrödinger equation
A wave equation that describes the behaviour of electrons (and other "particles") in quantum mechanical terms.

Strong nuclear force
Holds protons and neutrons together to make nuclei.

superposition of states
State of a quantum system when its wave function is made up of a mixture of waves corresponding to different physical possibilities.

thought experiment
An imaginary experiment, only carried out "in the mind", designed to demonstrate logically some point of view. Schrödinger's Cat Paradox is a classic example.

ultraviolet catastrophe
Prediction from classical mechanics that a hot object should radiate vast amounts of energy at short wavelengths, whatever its temperature. This is a catastrophe for classical theory, pointing the way to quantum physics, because it does not match the way black bodies radiate.

virtual particles
The particles exchanged between "real" particles during interactions – for example, the photons that carry the electromagnetic interaction between two electrons. They only exist while carrying the force.

w particles
Two of the intermediate vector bosons (see also z particle).

wave function
Equation describing the properties of a quantum entity, such as an electron, in terms of waves.

wave mechanics
A version of quantum mechanics based on the wave equation discovered by Erwin Schrödinger.

weak interaction
The "force" responsible for some kinds of radio-active decay.

x-rays
Electromagnetic radiation with wavelengths in the range from 12×10^{-9} to 12×10^{-12}m, slightly longer than gamma rays.

Young's slit experiment
Another name for the double-slit experiment, after Thomas Young, who carried out the experiment with light at the end of the 18th century.

z particle
One of the intermediate vector bosons (see also w particles).

index

Further reading

The Meaning of Quantum Theory, Jim Baggott, OUP, 1992

QED: The Strange Theory of Light and Matter, Richard Feynman, Penguin, 1990

Six Easy Pieces, Richard Feynman, Penguin 1995

Thirty Years that Shook Physics, George Gamow, Dover, 1966

In Search of Schrödinger's Cat, John Gribbin, Black Swan, 1984

Chaos & Uncertainty, Mary & John Gribbin, Hodder, 1999

The Cosmic Code, Heinz Pagels, Michael Joseph, 1982

Quantum Physics, Alastair Rae, CUP, 1986

Catching the Light, Arthur Zajonc, Bantam, 1993

....

Acknowledgments

Index
Indexing Specialists, Hove

Jacket design
Nathalie Godwin

Picture research
Lindsey Johns

Picture credits
Allsport: 7(tr). **Brand X Pictures:** 34(bl). **CERN:** 50(bl Higgs Boson), 51(tl), 51(tr). **Imperial War Museum:** 58(tl). **IBM Research, Almaden Research Center:** 26(l), 54(br) Unauthorized use not permitted. **NASA:** 58(bl). **Science Museum:** 17(b). **Science Photo Library:** American Institute of Physics 15(br), 31(tl), 17(tl), 18(br); American Institute of Physics/Physics Today Collection 37(cr); American Institute of Physics/Segre Collection 34(tl), 36(cl); American Institute of Physics/Sky and Telescope 25(tl), 30(tl); Clive Freeman, Biosym Technologies 43(tr), 43(cr); Tim Maylon 45(tr); Will and Deni McIntyre 46(bl); Hank Morgan 23(l); National Library of Medicine 8(br), 10(tl), 11(cl); David Parker (Peter Higgs portrait) 50(bl); Erich Schrempp 5(l), 6(cl); SLAC 47(cr); Volker Steger 54(tl quantum computer); Andrew Syred 55(tr); US Department of Energy 41(tr); Victor Habbick Visions 65(br); John Walsh 53(l). **Neil Setchfield (DK):** 38(cl). **SLAC (Courtesy of Stanford Linear Acceleration Centre):** 47(br), 48(tl).

Every effort has been made to trace the copyright holders.
The publisher apologizes for any unintentional omissions and would be pleased, in such cases, to place an acknowledgment in future editions of this book.

All other images © Dorling Kindersley.
For further information see: **www.dkimages.com**